Steve Jobs
Computer Pioneer & Co-Founder of Apple

by Grace Hansen

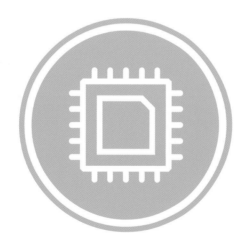

Abdo
HISTORY MAKER BIOGRAPHIES
Kids

abdobooks.com

Published by Abdo Kids, a division of ABDO, P.O. Box 398166, Minneapolis, Minnesota 55439.
Copyright © 2020 by Abdo Consulting Group, Inc. International copyrights reserved in all countries.
No part of this book may be reproduced in any form without written permission from the publisher.
Abdo Kids Jumbo™ is a trademark and logo of Abdo Kids.

Printed in the United States of America, North Mankato, Minnesota.

102019

012020

 THIS BOOK CONTAINS
RECYCLED MATERIALS

Photo Credits: Alamy, AP Images, Getty Images, iStock, newscom,
Seth Poppel/Yearbook Library, Shutterstock

Production Contributors: Teddy Borth, Jennie Forsberg, Grace Hansen
Design Contributors: Dorothy Toth, Pakou Moua

Library of Congress Control Number: 2019941231
Publisher's Cataloging-in-Publication Data

Names: Hansen, Grace, author.

Title: Steve Jobs / by Grace Hansen

Other title: Computer pioneer & co-founder of Apple

Description: Minneapolis, Minnesota : Abdo Kids, 2020 | Series: History maker biographies | Includes
    online resources and index.

Identifiers: ISBN 9781532189036 (lib. bdg.) | ISBN 9781532189524 (ebook) | ISBN 9781098200503
    (Read-to-Me ebook)

Subjects: LCSH: Jobs, Steve, 1955-2011--Juvenile literature. | Apple Inc.--Juvenile literature. | Technology-
    -Juvenile literature. | Businessmen--Biography--Juvenile literature. | Pixar Animation Studios--
    Juvenile literature. | iPhone Operating System--Juvenile literature. | Computers--Juvenile literature.

Classification: DDC 621.390 [B]--dc23

Table of Contents

Early Years

Steven Paul Jobs was born in San Francisco, California, on February 24, 1955. Soon after, he was adopted by Paul and Clara Jobs. His love of electronics came from his father. The two would take things apart and put them back together.

California

As a child, Steve had a hard time focusing in school. His 4th grade teacher helped him a lot. He did so well that year, that he skipped the 5th grade.

Think Different

Steve was **bullied** in school.

He wanted to move and go to a new school. His parents bought a home in Los Altos, California. This is where Jobs would meet Steve Wozniak.

9

Apple Computers

In 1976, Jobs and Wozniak
started Apple Computer in the
family's garage. Jobs was just
21 years old. Wozniak **invented**
the Apple I computer.

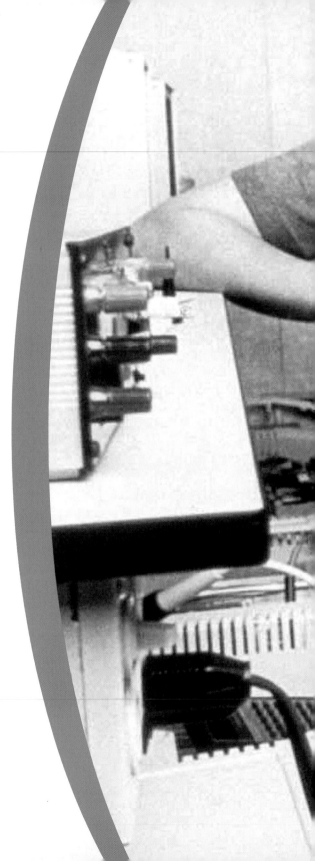

11

The two went on to make amazing personal computers. And they did it by making the machines smaller, cheaper, and simpler. But in 1985, Jobs left Apple. He began a new company called NeXT Inc.

Taking Apple to the Next Level

In 1996, Apple hired Jobs back. Just two years later, Jobs introduced the iMac computer.

In 2001, Jobs and Apple introduced iTunes. Later in the year, they launched the iPod. Now users could both purchase and play their music with Apple.

iPod
2001

In 2007, Jobs and Apple took the world by storm. This is when they introduced the iPhone. No one had ever seen anything like it before.

Death & Legacy

In 2003, Jobs was diagnosed with cancer. After years of illness, he died in 2011. In his life, he helped change the way people communicate and get information.

$1.50

A RECORD BUDGET
Battle of the Bulge

RICH

America's
Risk Takers

teven Jobs
f Apple Computer

KACZYNSKI SPEAKS

TIME

Steve's Jobs

He saved Apple with his hot new iMac. He struck gold at Pixar with digital movies like *Toy Story 2*. You'd think he'd learn to chill. Think different

www.time.com

AFGHANISTAN: DEADLY HUNT ■ INDIA & PAKISTAN: WAR DANCE

TIME

FLAT-OUT COOL!

Steve Jobs thinks he has seen the future—again. Apple's new iMac is an all-in-one hub for music, pictures and movies. It's elegant and affordable. But will millions of PC users get it?

TIME

WHAT'S NEXT

THE MAN WHO ALWAYS SEEMS TO KNOW...

Timeline

Paul and Clara move the family to Los Altos, California.

On June 5, the Apple II computer is released. It is Apple's first successful **personal computer**.

Jobs returns to Apple.

iTunes launches in January. In October, the first iPod **debuts**.

The world-changing iPhone is launched.

1968   **1977**   **1996**   **2001**   **2007**

1955   **1976**   **1985**   **1998**   **2011**

February 24
Steven Paul Jobs is born in San Francisco, California. He is adopted by Paul and Clara Jobs.

Steve Jobs and Steve Wozniak start Apple Computer in the family garage.

Jobs leaves Apple in September to start NeXT Inc.

In May, Jobs introduces the iMac computer.

October 5
At the age of 56, Jobs dies at home from **cancer** complications.

22

Glossary

bullied – harassed or frightened.

cancer – a disease in which certain cells divide and grow much faster than they normally do.

communicate – to exchange thoughts, ideas, or information.

debut – a first appearance.

invented – thought of and created something new.

personal computer – a computer designed for use by one person at a time.

Index

Abdo Kids
ONLINE
FREE! ONLINE MULTIMEDIA RESOURCES

Visit **abdokids.com**
to access crafts, games,
videos, and more!

Use Abdo Kids code

HSK9036

or scan this QR code!